ASQ Certified Quality Technician Practice Exams: 200 Practice Questions for the ASQ CQT Exam.

Study More Efficiently

Bova Books LLC
BovaBooks@gmail.com

Thank you for your Purchase!

We are available for full support of your studies for any questions or comments you may have. Email us at:

BovaBooks@gmail.com

Table of Contents

Exam Format and Expectations

Hello and thank you for your purchase. We are here for all of your studying needs on your way to passing the ASQ CQT exam. Please be sure to email us at BovaBooks@gmail.com for any questions you may have.

Our main goal when developing our study material is to focus your studying on the core concepts you need to know on test day. We develop our material exactly in line with the topics in the ASQ provided Body of Knowledge so you no longer have to waste time sieving through unnecessary material. Here is a breakdown of what to expect from the exam:

- The examination is multiple-choice and four hours long.

- Be sure to answer every question since there is no penalty for guessing.

- There are 110 questions on the exam

- 10 of the questions are not scored, however you will not know which ones are real so treat all of them as if they are real

- Exams are open book

Certified Quality Technician Syllabus

I. Quality Concepts and Tools (18 Questions)

    A. Quality Concepts

        1. Customers and suppliers Define internal and external customers, identify their expectations, and determine their satisfaction levels. Define internal and external suppliers and key elements of relations with them. (Understand)

        2. Quality principles for products and processesExplain basic quality principles related to products (such as features, fitness-for-use, and freedom from defects) and processes (such as monitoring, measuring, and continuous improvement). (Understand)

        3. Quality standards,requirements, and specifications Define and distinguish between national or international standards, customer requirements, and product or process specifications. (Understand)

        4. Cost of quality (COQ)Describe and distinguish between the four classic cost of quality categories (prevention, appraisal, internal failure, external failure) and classify activities appropriately. (Apply)

    B. Quality Tools The seven basic quality toolsSelect, construct, and interpret:

        1. Cause and effect diagrams (Evaluate)

        2. Flowcharts (process maps) (Evaluate)

        3. Check sheets (Evaluate)

        4. Pareto charts (Evaluate)

        5. Scatter diagrams (Evaluate)

        6. Control charts (Evaluate)

        7. Histograms (Evaluate)Topics in this body of knowledge (BoK) include additional detail in the form of subtext explanations and the cognitive level at which the questions will be written. This information will provide useful guidance for both the Exam Development Committee and the candidate preparing to take the exam. The subtext is not intended to limit the subject matter or be all-inclusive of what might be covered in an exam. It is meant to clarify the type of content to be included in the exam. The descriptor in parentheses at the end of each entry refers to the maximum cognitive level at which the topic will be tested. A complete description of cognitive levels is provided at the end of this document. Note: Approximately 20 percent of the questions in each CQT exam will require calculation.BODY OF KNOWLEDGECertified Quality Technician (CQT)

        8. Problem-solving techniquesDefine, describe, and apply problem solving techniques such as 5 Whys and 8D. (Apply)

        9. Six SigmaIdentify key Six Sigma concepts and tools such as quality function deployment (QFD), design of experiments (DOE), and design, measure, analyze, improve, control (DMAIC). (Remember)

        10. LeanIdentify key lean concepts and tools such as 5S, value-stream mapping, flow, and pull system. (Remember)

        11. Continuous improvement techniquesDefine and use various continuous improvement techniques including the plan-do-check-act (PDCA) cycle, brainstorming, and benchmarking. (Apply)

C. ASQ Code of Ethics for Professional Conduct Determine and apply appropriate behaviors and action that comply with this ethical code. (Evaluate)

II. Statistical Techniques   (17 Questions)
  A. General Concepts
      1. Terminology Identify and differentiate between statistical terms such as population, sample, parameter, statistic, and statistical process control (SPC). (Understand)
      2. Frequency distributions Define and compare normal, Poisson, and binomial frequency distributions. (Understand)
  B. Calculations
      1. Measures of central tendency Define, compute, and interpret mean, median, and mode. (Analyze)
      2. Measures of dispersion Define, compute, and interpret standard deviation, range, and variance. (Analyze)
      3. Confidence levels Explain confidence levels in various situations. (Understand)
      4. Confidence limits Explain confidence limits in various situations. (Understand)
      5. Probability Explain probability using the basic concepts of combinations, permutations, and area under the normal curve. (Understand)
  C. Control Charts
      1. Control limits vs. specification limits Identify and distinguish the different uses of control limits and specification limits. (Analyze)
      2. Variables charts Identify, select, construct, and interpret variables charts such as X-R and X-s. (Analyze)
      3. Attributes charts Identify, select, construct, and interpret attributes charts such as p, np, c, and u. (Analyze)
      4. Process capability measures Define the prerequisites for capability, and calculate and interpret capability indices (e.g., Cp, Cpk, Pp, Ppk) and capability ratio (CR) in various situations. (Analyze)
      5. Common and special cause variation Interpret various control chart patterns (e.g., runs, hugging, trends) and use rules for determining statistical control to distinguish between common cause and special cause variation. (Analyze)
8Certified Quality Technician
      6. Data plotting Identify the advantages and limitations of using this method to analyze data visually. (Understand)

III. Metrology and Calibration (18 Questions)
  A. Types of Measurement and Test Equipment (M&TE) Describe, select, and use the following types of M&TE, and evaluate their measurement results to determine conformance to specifications. (Evaluate)
      1. Hand tools (e.g., calipers, micrometers, linear scales, analog, digital, vernier scales)
      2.   Gauges (e.g., pins, thread, custom gauges, gage blocks)

3. Optical tools (e.g., comparators, profiles, microscopes)
4. Coordinate measuring machines (CMM) (e.g., touch probes, vision, laser)
5. Electronic measuring equipment (e.g., digital displays, output)
6. Weights, balances, and scales
7. Hardness testing equipment (e.g., Brinell, Rockwell)
8. Surface plate methods and equipment
9. Surface analyzers (e.g., profilometers, roughness reference standards)
10. Force measurement tools (e.g., torque wrenches, tensometers)
11. Angle measurement tools (e.g.,protractors, sine bars, angle blocks)
12. Color measurement tools (e.g., spectrophotometer, color guides, light boxes)
13. Automated in-line inspection methods (e.g., vision systems, laser inspection systems, pyrometers)

B. Control and Maintenance of M&TE

1. M&TE identification, control, and maintenance Describe various methodologies for identifying and controlling M&TE to meet traceability requirements, and apply appropriate techniques for maintaining such equipment to obtain optimum performance. (Apply)

2. Customer-supplied M&TE Describe and apply requirements for validation and control of customer-supplied equipment. (Apply) C. Calibration of M&TE 1. Calibration intervals Apply calibration schedules on the basis of M&TE usage history and risk. (Apply)2. Calibration results Interpret calibration results and the potential impact of using out-of-calibration tools or failing to calibrate equipment on a regular basis. (Analyze)

3. Calibration errorIdentify the causes of calibration error and its effect on processes and products. (Understand)

4. Hierarchy of standardsExplain the levels of standards (e.g., reference, primary, transfer) and their relationship to one another. (Apply)

IV. Inspection and Test (23 Questions)

A. Blueprint Reading and Interpretation

1. Blueprint symbols and components Interpret drawings and apply requirements in various test and inspection activities. (Analyze)

2. Geometric dimensioning and tolerancing (GD&T) Define and apply GD&T covered in the ASME Y14.5 standard. (Analyze)

3. Classification of product defect characteristics Define and distinguish between defect characteristics (e.g., critical, major, minor). (Analyze)

9Certified Quality Technician

B. Inspection Concepts

1. Types of measurements Define and select between direct, differential, and transfer measurements. (Understand)

2. Gauge selection Determine which measurement instrument to use considering factors such as resolution, accuracy, tolerance, environment, and product features. (Evaluate)

3. Measurement systems analysis (MSA) Define and distinguish between measurement terms such as correlation, bias, linearity, precision-to-tolerance, and percent agreement. Describe how gauge repeatability and reproducibility (R&R) studies are performed and how they are applied in support of MSA. (Analyze)

4. Rounding rules Use truncation and rounding rules on both positive and negative numbers. (Apply)

5. Conversion of measurements Convert between metric and English units. (Apply)

6. Inspection points Define and distinguish between inspection point functions (e.g., receiving, in-process, final, source, first-article), and determine what type of inspection is appropriate at different stages of production, from raw materials through finished product. (Analyze)

7. Inspection error Explain various types of inspection error, including operator error (e.g., parallax, fatigue), environment (e.g., vibration, humidity, temperature), and equipment (e.g., limitations, capability, setup). (Understand)

8. Product traceability Explain the requirements for documenting and preserving the identity of a product and its origins. (Apply)

9. Certificates of compliance (COC) and analysis (COA) Define and compare these two types of certificates. (Understand)

C. Inspection Techniques and Processes

1. Nondestructive testing (NDT) techniques Explain various NDT techniques (e.g., X-ray, eddy current, ultrasonic, liquid penetrant, magnetic particle). (Understand)

2. Destructive testing techniques Explain various destructive tests (e.g., tensile, fatigue, flammability). (Understand)

3. Other testing techniques Describe characteristics of testing techniques used for electrical measurement (e.g., DC, AC, resistance, capacitance, continuity), chemical analysis (e.g., pH, conductivity, chromatography), physical/mechanical measurement (e.g., hardness, pressure tests, vacuum, flow), and other techniques such as gravimetric testing, cleanliness testing, contamination testing, and environmental testing (e.g., bioburden, surface, air, water testing). (Remember)

D. Sampling

1. Sampling characteristics Identify and define sampling characteristics such as operating characteristic (OC) curve, lot size, sample size, acceptance number, and switching rules. (Apply)

2. Sampling types Define and distinguish between sampling types such as fixed sampling, single, double, skip lot, 100 percent inspection, attributes, and variables sampling. (Apply)

3. Selecting samples from lots Determine sample size (e.g., AQL), selection method and accept/reject criteria used in various situations. (Apply)

E. Nonconforming Material

1. Identifying and segregating Determine whether products or material meet conformance requirements, and use various methods to label and segregate nonconforming materials. (Evaluate)

2.   Material review process Explain various elements of this process such as the function of the material review board (MRB), the steps in determining fitness-for-use, and product disposition. (Understand)

V.   Quality Audits (12 Questions)
   A.   Audit Types and Terminology Define basic audit types:
      1) internal,
      2) external,
      3) systems,
      4) product,
      5) process. Distinguish between first-,second-, and third-party audits.
   (Understand)
   B.   Audit Components Describe and apply various elements of the audit process:
      1) audit purpose and scope,
      2) audit reference standard,
      3) audit plan (preparation),
      4) audit performance,
      5) opening and closing meetings,
      6) final report and verification of corrective action. (Apply)
   C.   Audit Tools and Techniques Define and apply various auditing tools:
      1) checklists and working papers,
      2) data gathering and objective evidence,
      3) forward- and backward-tracing,
      4) audit sampling plans and procedural guidelines. (Apply)
   D.   Audit Communication Tools Identify and use appropriate interviewing techniques and listening skills in various audit situations, and develop and use graphs, charts, diagrams, and other aids in support of written and oral presentations. (Apply)

VI.   Risk Management (12 Questions)
   A.   Risk Assessment and Mitigation Describe methods of risk assessment and mitigation such as trend analysis (SPC), failure mode and effects analysis (FMEA), root cause analysis (RCA), product and process monitoring reports, and control plans. (Understand)
   B.   Corrective Action Explain and apply elements of the corrective action process: identify the problem, contain the problem (interim action), assign responsibility (personnel) to determine the causes of the problem and propose solutions to eliminate it or prevent its recurrence (permanent action), verify that the solutions are implemented, and confirm their effectiveness (validation). (Apply)
   C.   Preventive Action Explain and apply elements of a preventive action process: use various data analysis techniques to identify potential failures, defects, or process deficiencies; assign responsibility for improving the process (e.g., develop error- or mistake-proofing devices or methods, initiate procedural changes), and verify the effectiveness of the preventive action. (Apply)

Question 1

Of the entities listed below, which can be classified as an internal customer?

(A) Insurance Company
(B) Advertising agency
(C) Human resources
(D) Raw materials supplier

Question 2

A specific concrete beam for a parking garage is found to be too long. The plans indicate the correct length of the beam. They are fabricated by a worker shifting the length of the form to the correct distance. What is most likely the source of variation category for the improper length?

(A) Machine
(B) Methods
(C) Materials
(D) Measurement

Question 3

Scrap material can be classified into which of the following cost-of-quality categories?

(A) Prevention
(B) Appraisal
(C) Internal failure
(D) External failure

Question 4

A product is being evaluated for cost efficiency. The company is evaluating ways of reducing costs. It is determined that 30% of material is scrap and can be reduced to save 2.1 million. The company will also reduce the frequency of calibration to save 0.4 million. Lastly the company plans to take action to reduce recall costs by 0.8 million. What is the total failure cost reduction?

(A) 0.8 million
(B) 1.2 million
(C) 2.9 million
(D) 3.3 million

Question 5

Which of the following is not one of the six most common categories of cause for a cause and effect diagram?

(A) Manipulation
(B) Measurement
(C) Machines
(D) Mother Nature

Question 6

What symbol is most often used to represent a decision point on a flow chart?

(A) Arrow
(B) Diamond
(C) Rectangle
(D) Circle

Question 7

An experiment is being designed to test concrete samples for strength. Eight samples are taken with the same amount of cement and aggregate but varying water content. The strength is then measured by loading each sample to failure. What is the dependent variable of the experiment?

(A) Cement
(B) Concrete
(C) Water
(D) Strength

Question 8

What statement below regarding experimental error is true?

(A) As replications increase the experimental error accuracy increases
(B) As replications decrease the experimental error accuracy increases
(C) As replications increase the experimental error accuracy decreases
(D) There is no relationship between replications and experimental error

Question 9

The chart below shows results from impact tests to measure the deflection of roadside barriers. The weight and speed of the truck is changed, and the corresponding deflection is measured. Determine the main effect of the weight factor from the data.

Weight (Tons)	Speed (MPH)	Deflection
2	40	2.0'
2	45	2.1'
2	50	2.4'
4	40	4.2'
4	45	4.3'
4	50	4.6'

(A) 1.1
(B) 1.8
(C) 2.5
(D) 4.6

Question 10

Which of the following distribution types are not considered discrete distributions?

(A) Binomial
(B) Poisson
(C) Standard normal
(D) All of the above are discrete distributions

Question 11

The interaction chart below shows three variables and an indication of high or low level for the given run. Determine the total number of positive interactions for AxB and AxC.

Run	A	B	C
1	+	+	+
2	+	+	-
3	+	-	-
4	-	+	+
5	-	-	+
6	-	-	-

(A) 2
(B) 4
(C) 6
(D) 10

Question 12

Which of the following is not considered a potential effect of implementing a fractional factorial design?

(A) Confounded factors
(B) Increased procedure time
(C) Lower quality results
(D) Lower material costs

Question 13

What is most often not considered a key element of a problem definition?

(A) Analysis results
(B) Problem statement
(C) Quality cost
(D) Criteria for solution

Question 14

Which of the following is not one of the five terms associated with the 5S tool?

(A) Sort
(B) Standardize
(C) Scrub
(D) Synthesize

Question 15

Which of the following accurately describes the completion of implementing the Plan-Do-Check-Act (PDCA) cycle?

(A) Action taken based on conclusions
(B) Review of responses to action taken
(C) Gathering of feedback
(D) The cycle does not complete

Question 16

The act of observing best practices of another organization to learn optimized techniques and strategies is called:

(A) Benchmarking
(B) Brainstorming
(C) Standardizing
(D) Pull system

Question 17

Which of the following tools allows for the incorporation of time into the representation of the data?

(A) Pareto chart
(B) Scatter plot
(C) Histogram
(D) Control chart

Question 18

A railroad project includes replacing rail ties for a 1000' length of track. 255 ties arrive on site and 5 of them are taken for testing. The term that best describes the group of 5 ties taken for testing is:

(A) Population
(B) Control
(C) Sample
(D) Universe

Question 19

What statement below is false regarding a standard normal distribution curve?

(A) The curve is symmetric about zero
(B) The mean is one unit divided by 2
(C) The curve never touches the x-axis
(D) The standard deviation is 1

Question 20

A shipment of timber planks for a deck is determined to have 1.2 unusable pieces per bundle. What is the probability of receiving a bundle with 2 defective pieces?

(A) 10%
(B) 22%
(C) 40%
(D) 50%

Question 21

There are 6 horses in an upcoming race. What is the number of combinations for any three horses from this particular race?

(A) 15
(B) 18
(C) 20
(D) 120

Question 22

For a binomial distribution which has 10 trials and a probability of success of 0.25, what is the mean?

(A) 2.5
(B) 5.6
(C) 8.8
(D) 12.4

Question 23

Which of the following statistical values is not related to a measurement of the dispersion of a data set?

(A) Range
(B) Mode
(C) Variance
(D) Standard deviation

Question 24

Concrete is tested for strength from two different producers. Producer A has a lower standard deviation than producer B. Which of the following statements is accurate?

(A) Strength of A is more reliable and has less variation than B
(B) Strength of B is more reliable and less variation Than A
(C) Strength of A is less reliable and less variation Than A
(D) Strength of B is less reliable and less variation Than A

Question 25

For a confidence level of 90%, what is the two-tailed confidence coefficient ($\alpha/2$)?

(A) 0.01
(B) 0.05
(C) 1.645
(D) 1.96

Question 26

Observations of attendance at a baseball stadium are made once per game for 162 games. The mean is determined to be 15,400 people and the 90% margin of error is 260. What is the upper and lower confidence limits?

(A) 15,140/15,660
(B) 15,140/15,400
(C) 13,860/16,940
(D) 15,140/16,940

Question 27

Which of the following is a true statement regarding specification limits?

(A) Indication of the variation of values based on data
(B) Cannot be used to identify a process change
(C) Designer target values
(D) Average values

Question 28

Which of the following is not a common indicator of a process change?

(A) A point above the UCL
(B) Seven successive points above the mean
(C) A group of 4 above the mean the 4 below the mean
(D) A point below the LCL

Question 29

Which of the following chart types is most appropriate for counting defectives with a constant sample size?

(A) p
(B) np
(C) u
(D) c

Question 30

A u-chart has the following data for defects and sample sizes:

Defects	3	2	3	4	5
Sample	60	45	51	72	71
Fraction	0.050	0.044	0.059	0.056	0.070

Determine the upper control limit.

(A) 0.042
(B) 0.058
(C) 0.099
(D) 0.147

Question 31

Which of the following relationships are true for processes that are not centered?

(A) $C_P = C_{PK}$
(B) $C_P < C_{Pk}$
(C) $C_P > C_{PK}$
(D) There is no correlation

Question 32

Which of the following metrology technologies mostly include contact instruments?

(A) Mechanical
(B) Air
(C) Light waves
(D) Electrons

Question 33

Which of the following can be classified as an error rather than systematic problem?

(A) Tape measure readings on lumber
(B) A scale calibrated to 0.01 lbs above zero
(C) A worn-out ruler difficult to read
(D) A warped wall of a concrete farm

Question 34

The graph shown below has data points on a scatter plot of distances measured (y-axis) for golf shots (x-axis). The target was at a distance of 200 yards away. What statement best describes the data?

(A) Precise and accurate
(B) Precise but not accurate
(C) Accurate but precise
(D) Neither accurate nor precise

Question 35

What is not classified as a linear measuring instrument?

(A) Steel rule
(B) Sine bar
(C) Vernier caliper
(D) Micrometer caliper

Question 36

The classification group of gauges used for the checking of dimensions during the production of a product is:

(A) Inspection
(B) Working
(C) Reference
(D) Master

Question 37

A plug gauge is used to measure a hole on a steel plate. The company uses a 20% working gauge tolerance and the part tolerance is 0.05 mm. What is the working tolerance of the measurement?

(A) 0.01 mm
(B) 0.02 mm
(C) 0.10 mm
(D) 0.20 mm

Question 38

Fitness-for-use of a product mostly refers to which of the following?

(A) Conforming to industry standard specifications
(B) Conforming to code and regulation standards
(C) Providing the customer a product they will want and actually use
(D) Providing the customer a product they didn't know they wanted

Question 39

Of the inspection needs listed below which is most likely the least appropriate use of a template?

(A) Inspection of a groove
(B) Inspection of an outside radius tangent to two perpendicular planes
(C) Inspection of a ridge segment
(D) Inspection of and inner pipe dimeter

Question 40

Which of the following gauge block types is least appropriate for use in a highly corrosive environment?

(A) Ceramic
(B) Steel
(C) Tungsten-carbide
(D) Chrome-carbide

Question 41

Which of the following is not a type of air pressure gauge?

(A) Back-pressure
(B) Electrical-mechanical
(C) Venturi
(D) Differential

Question 42

What hardness test is performed by using a carbide ball to apply a force to a material causing an indentation which is measures?

(A) Knoop
(B) Vickers
(C) Brinell
(D) Rockwell

Question 43

What American National Standards Institute (ANSI) symbol represents the lay approximately parallel to the line representing the surface to which the symbol is applied?

(A) =
(B) X
(C) M
(D) P

Question 44

The ten-point height is obtained by averaging which of the following values within the sampling length relative to a straight mean line?

(A) The ten highest points
(B) The ten lowest points
(C) The 5 highest points and 5 lowest points
(D) The average of the high and low points

Question 45

When M&TE is customer supplied, which of the following statements does not meet the requirements of ISO/IEC 17025?

(A) A unit shall be taken out of service based on field decisions of wear and tear
(B) Calibration status shall be indicated on the instrument
(C) Calibration schedules must be kept
(D) Equipment shall be identified and documented

Question 46

An acceptable calibration interval may be any of the following except:

(A) Weekly
(B) Every 10,000 uses
(C) Every 10 months based on historical performance
(D) All of the above

Question 47

Which of the following standards has the highest precedence related to the hierarchy of calibration standards?

(A) National
(B) Working
(C) Primary
(D) International

Question 48

In regards to maximum and least material condition for dimensional tolerance, which of the following statements are false?

(A) The MMC of a shaft would be the maximum diameter
(B) The MMC of a hole would be the minimum diameter
(C) The LMC of a shaft would be the maximum diameter
(D) The LMC of a hole would be the maximum diameter

Question 49

Which of the following measurement instruments can be classified as indirect?

(A) Gauge block
(B) Steel rule
(C) Optical flat
(D) Caliper

Question 50

The tolerance on a steel rod diameter is determined to be 0.25". Using the common practice of gauge maker's rule (rule of ten), The increment of measurement on the instrument for inspection shall be no greater than:

(A) 0.01"
(B) 0.025"
(C) 0.1"
(D) 0.25"

Question 51

Determine the test uncertainty ratio for a digital caliper with an upper and lower specification limit of 0.012 and 0.008 respectively. The estimate uncertainty is 0.0002. Determine the test uncertainty ratio.

(A) 1.2
(B) 4.4
(C) 5.6
(D) 10.0

Question 52

Determine the mass (kg) of a person who weighs 110 lbs.

(A) 22
(B) 50
(C) 120
(D) 242

Question 53

Data from a test for a gauge resulted in a repeatability of 7.25 and a reproducibility of 0.75. Determine the gauge R&R.

(A) 4.4
(B) 5.2
(C) 7.3
(D) 11.3

Question 54

A company has a standard inspection procedure for a specific product to inspect on a daily basis only the first few products of the process and then allow the rest to be made without any further inspection. What type of inspection is being performed?

(A) Pre-process
(B) Final
(C) Verification
(D) Incoming material

Question 55

The tracking of the movement of products in multiple processes between manufacturers is which of the following?

(A) Chain traceability
(B) Internal traceability
(C) Extraction tracing
(D) Parallax

Question 56

The Certificate of Analysis (COA) is mostly identified as proving compliance with which of the following?

(A) Local code requirements
(B) Specification requirements
(C) Customer requirements
(D) National Code requirements

Question 57

What nondestructive testing technique is least suited for the detection of surface defects?

(A) Ultrasonic
(B) Liquid penetration
(C) X-ray
(D) Dye penetrant

Question 58

What sampling type is not allowed by ANSI/ASQ Z1.4-2008?

(A) Selective sampling
(B) Single sampling
(C) Double sampling
(D) Multiple sampling

Question 59

Calculate the average outgoing quality for a sample size of 20 in a lot of 200. The probability of acceptance is 0.954 and the nonconforming fraction is 0.01.

(A) 0.0086
(B) 0.0154
(C) 0.0250
(D) 0.10

Question 60

What primary quality audit type most often has the largest scope?

(A) Process
(B) Product
(C) System
(D) General

Question 61

What statement is best associated with consumer's risk as opposed to producer's risk?

(A) The probability of rejecting a lot that is actually within AQL requirements
(B) The probability of rejecting a lot that is actually outside of AQL requirements
(C) The probability of rejecting a lot that has a quality level equal to LTPD
(D) The probability of rejecting a lot that has a quality level beyond LTPD

Question 62

What is the most common probability of acceptance to determine lot tolerance percent defective for a sampling plan?

(A) 2%
(B) 10%
(C) 50%
(D) 75%

Question 63

What is least likely a function of the Material Review Board (MRB)?

(A) Review non-conforming products
(B) Determine material disposition
(C) Investigate non-conforming root causes
(D) Segregate non-conforming products

Question 64

Which of the following is not one of the basic components of an audit?

(A) Scoping
(B) Performance
(C) Reiterations
(D) Closure

Question 65

Which of the following audit types is correlated to an internal audit source?

(A) First-party
(B) Second-party
(C) Third-party
(D) Fourth-party

Question 66

What is not associated with the preparation component of audits?

(A) Identification of authorization source
(B) Define resources
(C) Define roles and responsibilities
(D) Begin collecting data

Question 67

Which of the following audit tools are not specifically beneficial for ensuring all process steps are completed?

(A) Checklist
(B) Sampling plans
(C) Forward and backward tracing
(D) Flowchart

Question 68

A failure mode and effect analysis (FMEA) has determined a potential failure to have a 2, 4, and 6 for severity, occurrence, and detection respectively. What is the risk priority number (RPN)?

(A) 48
(B) 67
(C) 100
(D) 408

Question 69

When evaluating potential failures with the same risk priority, what is the first action to be taken as per Palady (1997)?

(A) Eliminate occurrence
(B) Reduce occurrence
(C) Reduce severity
(D) Improve detection

Question 70

Which of the following is not an element associated with corrective action?

(A) Transfer risk
(B) Identify problem
(C) Containment
(D) Solution verification

Question 71

What is least likely an element associated with defining preventative action?

(A) Identify the potential problem
(B) Provide temporary solution
(C) Process implementation
(D) Verification

Question 72

A company has implemented a corrective action process and needs to verify its effectiveness. The action has a low opportunity for occurrence and observation. What statement is accurate in regard to the time needed after implementation?

(A) Increased time
(B) Decreased time
(C) No effect
(D) Unpredictable

Question 73

The investigation of a pathway by which something is negatively impacted by some hazard defines which type of root cause analysis?

(A) Five "whys"
(B) Barrier
(C) Change
(D) Failure mode and effect

Question 74

Which of the following scenarios is an example of preventative action?

(A) Fixing broken concrete decks after installation for a single occurrence
(B) Applying anti-corrosion coatings
(C) Replacement of computer systems
(D) Replacing machine parts that have been determined to cause skewed installations

Question 75

What element below is least likely to be appropriate for a cause and effect diagram?

(A) Problem statement
(B) Major categories of root problem
(C) Cost vs. time plot
(D) Minor categories of root problem

Question 76

What quality tool listed below is least appropriate for the use of stratification?

(A) Control chart
(B) Histogram
(C) Scatter plot
(D) Fishbone diagram

Question 77

Which PH reading below is the most acidic?

(A) 3
(B) 6
(C) 8
(D) 10

Question 78

Which of the following electrical measuring devices records voltages as a function of time?

(A) Oscilloscope
(B) Voltmeter
(C) Pulse generator
(D) LCR meter

Question 79

When determining if pipe diameters are less than a maximum allowable, a producer compares them to a master cylinder. What method of measurement is being used?

(A) Direct
(B) Indirect
(C) Comparative
(D) Transposition

Question 80

What index ratio below indicates the capability of a process where the mean is not centered between spec limits?

(A) C_P/P_P
(B) C_{Pl}/P_{Pl}
(C) C_{Pu}/P_{Pu}
(D) C_{Pk}/P_{Pk}

Question 81

Which of the following is an example of a cost associated with poor quality?

(A) Warranties
(B) Capability evaluations
(C) Product reviews
(D) Quality planning

Question 82

According to the Kano model, what is not one of the three levels of customer expectations?

(A) Expected
(B) Normal
(C) Relative
(D) Exciting

Question 83

What management and planning tools listed below organizes ideas from brainstorming into groups of related thoughts?

(A) Affinity diagram
(B) Tree diagram
(C) Interrelationship diagram
(D) Arrow diagram

Question 84

What is least likely a component of a house of quality diagram when implementing Quality Function Deployment?

(A) Voice of customer
(B) Specification Review
(C) Importance rating
(D) Target values

Question 85

A "Just-in-Time" production model can be characterized by which of the following?

(A) Products are overstocked to always be available
(B) Products must meet a minimum storage threshold
(C) Products are produced to meet actual demands
(D) Products are produced to meet forecasted demands

Question 86

A stable process may contain how many instances of special cause variation?

(A) 0
(B) 1
(C) 2
(D) 3

Question 87

According to Geometric Dimensioning and Tolerancing Standard ASME Y 14.5, what angle needs an explicitly shown angular dimension?

(A) 45°
(B) 90°
(C) 180°
(D) 270°

Question 88

According to Geometric Dimensioning and Tolerancing Standard ASME Y 14.5, what is the temperature (in degrees Fahrenheit) used as a baseline to establish dimensional tolerances?

(A) 60°
(B) 65°
(C) 68°
(D) 70°

Question 89

A gas station pump has an uncertainty of 0.003 gallons/gallon. If a customer purchases 16 gallons, what is the minimum amount of gas that may actually be received?

(A) 15.905
(B) 15.952
(C) 15.971
(D) 15.995

Question 90

Pyrometers determine the temperature as it relates to all of the following except:

(A) Thermal radiation
(B) Carbon content
(C) Constant of proportionality
(D) Emissivity

Question 91

Which of the following distribution types is a representation of only two potential outcomes?

(A) Normal
(B) Poisson
(C) Binomial
(D) Standard

Question 92

A switching plan may alternate between all of the following inspection types except:

(A) Normal
(B) Reduced
(C) Tightened
(D) Auxiliary

Question 93

Which of the following sample plan types includes taking the highest number of samples?

(A) Single
(B) Double
(C) Sequential
(D) Multiple

Question 94

Which of the following is true regarding multiple sampling?

(A) A lot must be tested more than once
(B) A lot must be tested the specified number of successive samples
(C) A lot may be accepted after the first test
(D) A lot can only be accepted after at least two tests

Question 95

A company determines the following costs (in millions of dollars) from an internal audit. What is the total cost of quality?

Prevention	4
Appraisal	8
Internal Failure	9
External Failure	6

(A) 12
(B) 18
(C) 27
(D) 30

Question 96

What statement below regarding the differences between internal and external customers is false?

(A) Internal customers are associated with the organization
(B) External customers can be the end user
(C) Internal customers may benefit from the sale of product
(D) Internal customers may not be end users

Question 97

Which perspective of quality is mostly defined as products being done well above the minimum expectations?

(A) Customer view
(B) Producer view
(C) Transcendent view
(D) Supplier view

Question 98

What statement below is mostly associated with quality control (QC) rather than quality assurance (QA)?

(A) Focus on product
(B) Focus on process
(C) Focus on defect prevention
(D) Focus on verification

Question 99

What variable types are associated with Statistical Quality Control (SQC)?

(A) Independent
(B) Dependent
(C) Process
(D) Input

Question 100

What is the determining factor when deciding to use an R or an S chart?

(A) System stability
(B) Control limits
(C) Number of subgroups
(D) Mean over time

Question 101

What is not one of the three project constraints?

(A) Cost
(B) Process
(C) Time
(D) Scope

Question 102

Which of the following actions is most likely to affect a products fitness-for-use?

(A) Increasing production efficiency
(B) Product changes resulting from consumer questionnaires
(C) Reduction in cost which results in reduction in quality
(D) Increasing defect tolerance

Question 103

Which of the following is not included in the calculation of total failure costs?

(A) Internal failure
(B) External failure
(C) Appraisal
(D) All the above are included in total failure costs

Question 104

What is not a step in the 8D problem solving method?

(A) Form a team
(B) Determine root cause
(C) Implement corrective actions
(D) Devalue the impact of the problem

Question 105

The most common initial step in the development of a Quality Function Deployment Matrix is:

(A) Identify technical requirements
(B) Identify customer requirements
(C) Determine Target values
(D) Action notes

Question 106

An experiment is being designed to test concrete samples for strength. Eight samples are taken with the same amount of cement and aggregate but varying water content. The strength is then measured by loading each sample to failure. What is the independent variable of the experiment?

(A) Cement
(B) Concrete
(C) Water
(D) Strength

Question 107

What statement below is true regarding the relationship between the calculated main effect and the influence of the effect on the quality characteristic?

(A) The lower the absolute value of the main effect, the greater the influence on the quality characteristic
(B) The higher the absolute value of the main effect, the greater the influence on the quality characteristic
(C) The higher the absolute value of the main effect, the less the influence on the quality characteristic
(D) There is no relationship between main effect and the quality characteristic

Question 108

The interaction chart below shows three variables and an indication of high or low level for the given run. Determine the total number of positive interactions for AxBxC.

Run	A	B	C
1	+	+	+
2	+	+	-
3	+	-	-
4	-	+	+
5	-	-	+
6	-	-	-

(A) 1
(B) 3
(C) 4
(D) 6

Question 109

An experiment is being designed to test fatigue life on steel members. Due to staffing restrictions the number of runs must be limited to no greater than 200. The steel will be tested at different strengths, thicknesses and temperatures. What is the maximum amount of levels that may be tested?

(A) 2
(B) 3
(C) 4
(D) 5

Question 110

Scope creep can best be defined as which of the following?

(A) Tendency for a team to expand the problem definition
(B) Increased costs in the estimate
(C) Protocols for addressing errors
(D) The tendency for materials to have inherent errors

Question 111

An employee who must wash down the surface of a conveyor belt once per hour needs to walk 250 ft each time to activate the machine for washing. This type of waste falls under which of the following categories?

(A) Waiting
(B) Defects
(C) Inventory
(D) Motion

Question 112

What system of production includes a traveling stocker who must be signaled to refill specific materials when they are running low?

(A) Pull system
(B) 5S
(C) Kanban
(D) Flow

Question 113

Which of the following can be classified as a hard problem-solving tool?

(A) Brainstorming
(B) Flow charts
(C) Cost comparisons
(D) Check sheets

Question 114

Of the actions listed below, which is not one of the main expectations of the ASQ code of ethics?

(A) Integrity and Honesty
(B) Safeguard proprietary information
(C) Avoid competitive advantages
(D) Avoid conflicts of interest

Question 115

What option below is most appropriate for identifying the most common source of failure for a particular product?

(A) Pareto chart
(B) Scatter plot
(C) Check sheet
(D) Control chart

Question 116

A coin is tossed 12 times. The probabilities of the coin being heads exactly one, two, or three are 0.0029, 0.0161, and 0.0537 respectively. What is the probability of getting one, two, or three heads in the 12 tosses?

(A) 0.0041
(B) 0.0552
(C) 0.0727
(D) 0.0954

Question 117

For a standard normal distribution curve with a unit of 0.01 in, determine the area under the curve.

(A) 0.001 in^2
(B) 0.01 in^2
(C) 0.02 in^2
(D) 1.0 in^2

Question 118

A car model has the possibility of 5 different defects that require a recall. For any given 100 cars, the average amount of cars with a defect is 1.2 and follows a Poisson distribution. What is the standard deviation?

(A) 0.553
(B) 0.955
(C) 1.095
(D) 2.130

Question 119

A set of data points has 10 values. How is the median of the set determined?

(A) The median is the 5th number
(B) The median is the 6th number
(C) The median is the average of all 10 pints
(D) The median is the average of the 5th and 6th number

Question 120

Which of the following statements is false regarding the mode of a data set?

(A) If no value occurs more than once, there is no mode
(B) If more than one value tie for number of appearances, the mode is the average of all tied numbers
(C) A data set can have more than one mode
(D) Modes are the highest bar on a histogram

Question 121

Which of the following equations is a representation of a population standard deviation?

(A) $s^2 = \frac{\Sigma(x-\bar{x})^2}{n-1}$

(B) $s = \sqrt{\frac{\Sigma(x-\bar{x})^2}{n-1}}$

(C) $\sigma = \sqrt{\frac{\Sigma(x-\mu)^2}{N}}$

(D) $s = \sqrt{\frac{\Sigma(x-\mu)^2}{N-1}}$

Question 122

For a confidence level of 95%, what is the normalized confidence coefficient?

(A) 0.05
(B) 0.025
(C) 1.96
(D) 2.575

Question 123

Determine the range from the following data set: 121, 190, 132, 145, 98, 185

(A) 92
(B) 98
(C) 121
(D) 150

Question 124

What best defines the variance of a data set?

(A) Average of the squared differences from the mean
(B) Difference between the highest and lowest value of a data set
(C) Square root of the average of the mean
(D) Square root of the standard deviation

Question 125

Three separate observations are taken on a process for three separate days. Defects are counted per hour each day over a 4-hour period. The results are shown below:

Day/Hour	1	2	3	4
1	11	21	14	11
2	6	11	12	13
3	18	4	10	6

Determine the process mean.

(A) 11.42
(B) 13.35
(C) 15.66
(D) 18.22

Question 126

The probability of a random number falling in a specific range for a standard normal distribution is equal to:

(A) The area under the curve in the specific range
(B) The area under the curve outside of the specified range
(C) One standard deviation outside of the range
(D) The total area under the curve minus the area under the range

Question 127

Which of the following best defines control limits?

(A) Indication of the variation of values based on data
(B) Limits based on customer feedback
(C) Designer target values
(D) Average values

Question 128

Which of the following chart types is most appropriate for counting defects with a variable sample size?

(A) p
(B) np
(C) u
(D) c

Question 129

An individuals and moving range chart is used when the sample size is:

(A) 1
(B) 5
(C) >25
(D) >50

Question 130

What type of profilometer makes contact with the analyzed surface?

(A) Optical
(B) Stylus
(C) Infrared
(D) 3-Dimensional

Question 131

For the control chart shown below, the UCL and LCL are 150 and 100 respectively. The average line is at 120 Which of the following indicators apply?

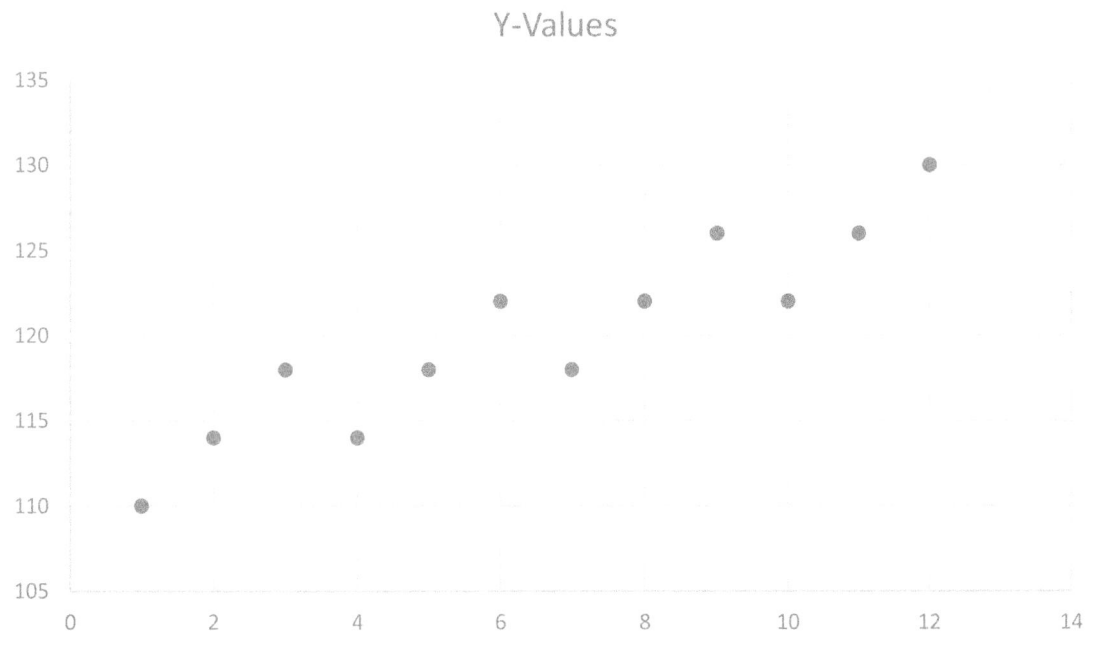

(A) 7 successive points above the average line
(B) 7 successive points trending
(C) Nonrandom patterns
(D) No process change indicators apply

Question 132

Calculate the average outgoing quality for a probability of acceptance of 0.981 and a nonconforming fraction of 0.02 if the lot size is considered to be infinitely large.

(A) 0.0088
(B) 0.0123
(C) 0.0196
(D) 0.0223

Question 133

What is not a general class of measurement?

(A) Legal
(B) Technical
(C) Predictive
(D) Scientific

Question 134

The capability ratio is equal to which of the following:

(A) $C_p/1$
(B) $1/C_p$
(C) $1 + C_p$
(D) $1 - C_p$

Question 135

A micrometer has 25 graduated thimble divisions. The lead for the thread on the spindle is 0.25 mm. What is the distance between each thimble graduation?

(A) 0.001 mm
(B) 0.01 mm
(C) 0.1 mm
(D) 0.25 mm

Question 136

Which of the following gauge types is most appropriate for checking if the inside diameter of a PVC pipe is acceptable?

(A) Plug
(B) Ring
(C) Snap
(D) Spline

Question 137

If the tolerance on a gauge is set so that the unilateral tolerance is outside of the work tolerance, what statement below is true?

(A) Some acceptable parts will be rejected
(B) Some unacceptable parts will be accepted
(C) Some of both rejected and accepted parts will be judged incorrectly
(D) There is no determination regarding acceptable/rejected products

Question 138

What is not an actual advantage of using a Coordinate Measuring Machine (CMM)?

(A) CMM's utilize multiple coordinate systems to provide more iterations
(B) Multiple measurements per setup
(C) Automatic data recording
(D) Ability to save data progress

Question 139

Which of the following gauge block standards has an accuracy of 0.00003 mm per 25 mm?

(A) Federal accuracy grade 0.5
(B) Federal accuracy grade 1.0
(C) Federal accuracy grade 1.5
(D) Federal accuracy grade 2.0

Question 140

Which of the following material types for gauge blocks are best suited for an environment that is not well controlled causing changes in temperature?

(A) Steel
(B) Ceramic
(C) Chrome-carbide
(D) Tungsten-carbide

Question 141

An architectural brick with a textured surface is to be measured using a multisensory coordinate measuring machine (MSCMM). Which of the following probe types is most appropriate for this type of measurement?

(A) Hard probe
(B) Touch trigger probe
(C) Optical probe
(D) Laser scanning probe

Question 142

The chart below shows surface profile measurements taken about an X-axis for a stone. Determine the average surface roughness for the measurements.

1	0.02
2	-0.03
3	-0.06
4	0.01
5	-0.04

(A) 0.010
(B) 0.023
(C) 0.032
(D) 0.044

Question 143

For optical flats, a dark band indicates a change of how much of a wavelength in distance separating the work surface and flat?

(A) 1/4
(B) 1/2
(C) 3/4
(D) 1

Question 144

The length of a sine bar is 5 inches. The bar is used to measure an angle and measurements are taken to be 0.75" and 4.25" at each end respectively. What is the measured angle?

(A) 22.6°
(B) 31.5°
(C) 38.7°
(D) 44.4°

Question 145

ISO 9001 standards requires all of the following in relation to measuring equipment except:

(A) Instruments shall be preadjusted as necessary
(B) Remove instruments where no such national standards exist
(C) Identify instruments to determine calibration status
(D) Take steps to protect from deterioration throughout the life of the instrument

Question 146

Which of the following represents a measurement with a unilateral tolerance?

(A) $4.00^{+0.00/-0.004}$
(B) $4.00^{+0.04/-0.004}$
(C) $4.00^{+0.05/-0.004}$
(D) $4.00^{-0.05/-0.004}$

Question 147

Which of the following is not a symbol where Maximum Material Condition (MMC) can be applied?

(A) Parallelism
(B) Straightness
(C) Roundness
(D) Angularity

Question 148

What symbol below indicates true position?

(A) ▱
(B) ⊕
(C) △
(D) ◠

Question 149

What defect classification is characterized by the potential for injury or significant economic loss?

(A) Critical
(B) Serious
(C) Major
(D) Minor

Question 150

Which of the following can be applied to the term defect but not nonconformity?

(A) Fitness-for-use of the product
(B) Failure to conform to specifications
(C) Diminishing product quality
(D) Defect and nonconformity are interchangeable entirely

Question 151

Measurements taken are to be rounded to 4 significant digits. A measurement in the field is taken as 45.67483". What is the correct rounded value?

(A) 45.7
(B) 45.67
(C) 45.68
(D) 45.6748

Question 152

Which of the following would not fall under pre-process inspection?

(A) Inspect in preparation for high cost activities
(B) Inspect products with a historically high rate for defects
(C) Inspection prior to painting
(D) inspect microchips after production but before installation

Question 153

What is the most common minimum accepted test accuracy ratio (TAR)?

(A) 1:1
(B) 2:1
(C) 4:1
(D) 10:1

Question 154

Which of the following nondestructive test methods is limited to conducting materials?

(A) Eddy current
(B) Ultrasonic
(C) Liquid penetration
(D) X-ray

Question 155

The Average outgoing quality limit is best defined as which of the following?

(A) Probability that a lot will be accepted
(B) Minimum amount of defective incoming product
(C) Maximum percentage of defective outgoing product
(D) Maximum percentage of incoming excess product

Question 156

As per ANSI/ASQ Z1.4-2008 what is not one of the seven levels of inspection?

(A) Heightened
(B) Reduced
(C) Tightened
(D) Normal

Question 157

Of the following options, which is not a standard sampling plan as per ANSI/ASQ Z1.9-2008?

(A) Variability known
(B) Variability known standard deviation
(C) Variability unknown standard deviation
(D) Variability unknown range method

Question 158

What is not a primary quality audit type?

(A) Process
(B) System
(C) Interpretive
(D) Product

Question 159

What statement below is the exception to the use of the special addition rule of probability?

(A) If A and B occur no more than 5 times each
(B) A and B may occur simultaneously
(C) A and B may not occur simultaneously
(D) A and B must not overlap

Question 160

For a double sampling plan, if the acceptance number for the second test is 4 defects, what is the rejection number?

(A) 5
(B) 6
(C) 7
(D) 8

Question 161

What statement is best associated with producer's risk as opposed to consumer's risk?

(A) The probability of rejecting a lot that is actually within AQL requirements
(B) The probability of rejecting a lot that is actually outside of AQL requirements
(C) The probability of rejecting a lot that has a quality level equal to LTPD
(D) The probability of rejecting a lot that has a quality level beyond LTPD

Question 162

What is not an acceptable action related to the segregation of non-conforming materials?

(A) Immediate re-insertion
(B) Initiate material review process
(C) Scrap
(D) Eventual reinsertion

Question 163

A bolt is manufactured to connect two plates but is fabricated ¼" too long. The Material Review Board determines that by the plans the bolt will still fit and can be used. Which of the following dispositions is most appropriate?

(A) Use as-is
(B) Scrap
(C) Repair
(D) Downgrade

Question 164

A state Department of Transportation is provided funds from a Federal agency for specific projects. If the agency performs an audit, which of the following audit types will it be?

(A) First-party
(B) Second-party
(C) Third-party
(D) Fourth-party

Question 165

Which of the activities below are not associated with the audit component of performance?

(A) Identify authorization source
(B) Collect data
(C) Exit meeting
(D) Analyzing data

Question 166

A data set collected from an internal audit is determined to be too large. What is not an appropriate means of reducing the data to a smaller sample?

(A) Reduced set must accurately characterize the larger
(B) Reduced set shall be no less than 50% of the larger set
(C) Biased data shall be removed
(D) Data must be selected randomly relative to its characteristics

Question 167

When using failure mode and effect analysis (FMEA) for risk assessment, what is not one of the components of determining a risk priority number?

(A) Frequency
(B) Severity
(C) Occurrence
(D) Detection

Question 168

What is the maximum risk priority number?

(A) 1
(B) 10
(C) 100
(D) 1000

Question 169

When evaluating risk using FMEA, which of the following statements are true?

(A) Only the actual failure may be rated
(B) Only the cause of failure may be rated
(C) Both the cause and actual failure may be rated
(D) Neither the cause nor the actual failure may be rated

Question 170

What is not a type of Failure Mode and Effect Analysis (FMEA)?

(A) System
(B) Design
(C) Integrated
(D) Process

Question 171

What corrective action element can be identified as interim action?

(A) Identify the problem
(B) Validate solutions
(C) Determine causes
(D) Containment of the problem

Question 172

What is not a method for verifying the effectiveness of a corrective action plan (CAP)?

(A) Audit
(B) Root cause evaluation
(C) Sampling
(D) Trend analysis

Question 173

The root cause analysis which identifies the limited number of tasks that produce a significant impact (80/20 rule) is which of the following?

(A) Pareto
(B) Fault tree
(C) Scientific
(D) Iterative analysis

Question 174

Which of the following scenarios is an example of corrective action?

(A) Fixing broken concrete decks after installation for a single occurrence
(B) Applying ant-corrosion coatings
(C) Replacement of computer systems
(D) Replacing machine parts that have been determined to cause skewed installations

Question 175

Which of the following is true of corrective action as oppose to a correction?

(A) Immediate response
(B) Reactive action
(C) Temporary
(D) Planned and investigative

Question 176

Which of the following is not an instrument geometry for the use of spectrometers?

(A) Single angle
(B) Variable
(C) Spherical
(D) Multi-angle

Question 177

Pure distilled water has a PH level of which of the following?

(A) 4
(B) 5
(C) 7
(D) 8

Question 178

What device listed below can be used to measure capacitance?

(A) Voltmeter
(B) Ohmmeter
(C) Ammeter
(D) LCR meter

Question 179

If a process is determined to have a capability index less than 1 ($C_P < 1$), which statement below is true of the process?

(A) The process spread has values that fall outside of the specification spread
(B) The process spread falls entirely within the spec spread
(C) The spec spread is much greater than the process spread
(D) The process is generally acceptable

Question 180

What index ratio below indicates the capability of a process where the mean meets the lower spec limits?

(A) C_P/P_P
(B) C_{Pl}/P_{Pl}
(C) C_{Pu}/P_{Pu}
(D) C_{Pk}/P_{Pk}

Question 181

Which of the following is an example of a cost associated with good quality?

(A) Warranties
(B) Repairs
(C) Error proofing
(D) Repeated services

Question 182

According to the Kano model, what level of customer expectations has the ability to dissatisfy a customer but not fully satisfy?

(A) Expected
(B) Normal
(C) Relative
(D) Exciting

Question 183

The main focus of the Quality Function Deployment is mostly:

(A) Process efficiency
(B) Cause and effect
(C) Responding to the voice of the customer
(D) Brainstorming

Question 184

What test below is a destructive testing technique?

(A) Liquid penetration
(B) Magnetic particle
(C) Ultrasonic
(D) Charpy V-notch

Question 185

What statement is mostly true in regard to a lean pull system?

(A) Products are based on actual demand
(B) Products are based on forecasted demands
(C) Both A and B
(D) Products are overstocked to ensure availability

Question 186

What example below is most likely a special cause variation?

(A) Use of a new cement type in a concrete mix
(B) Temperature causing elongations of steel members
(C) A product being within tolerance variations of plastic pipe diameters
(D) Production line quality control personnel errors

Question 187

What geometric characteristic below does not use a datum reference?

(A) Flatness
(B) Perpendicularity
(C) Angularity
(D) Position

Question 188

What is the standard minimum amount of time for the frequency of calibration?

(A) 1 month
(B) 6 months
(C) 1 year
(D) There is no standard minimum

Question 189

A voltage meter has an uncertainty of +/- 0.05 V with a 95% confidence. The range of values follows a normal distribution. If the meter displays a reading of 11 V, what is the percentage chance that the actual value is less than 10.95?

(A) 1%
(B) 2.5%
(C) 5%
(D) 10%

Question 190

What is likely the most appropriate choice for detecting defects on a continuous material such as paper?

(A) 1-D
(B) 2-D
(C) 3-D
(D) Area Scan

Question 191

What statement below is true regarding the relationship between the value of the capability ratio and how capable a process is?

(A) Higher CR implies higher capability
(B) Higher CR implies higher capability index
(C) Lower CR implies higher capability
(D) There is no clear connection

Question 192

Switching rules for a sampling plan are shown below:

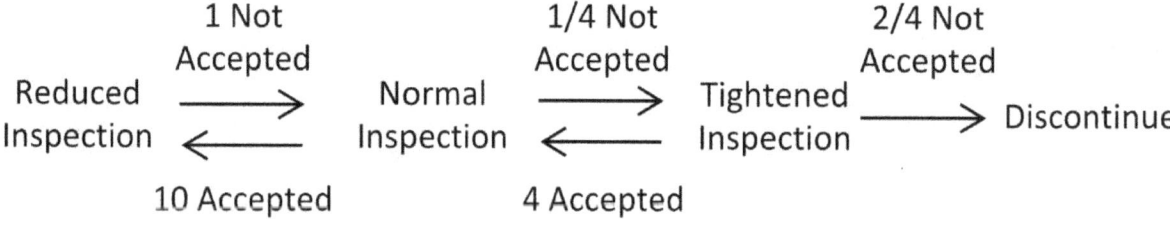

The following ratios of rejected:accepted came out: 0:10, 1:4, and 1:10. What inspection type is being used after the indicated ratios?

(A) Normal
(B) Reduced
(C) Tightened
(D) Tightened

Question 193

Which of the following sample plan types is most often the highest cost?

(A) Single
(B) Double
(C) Sequential
(D) Multiple

Question 194

The cost of good quality is the combination of appraisal cost and which of the following?

(A) Internal failure
(B) External failure
(C) Prevention
(D) Customer satisfaction

Question 195

Which of the following can be used to describe a special cause source of variation?

(A) Assignable
(B) Non-assignable
(C) Consistent
(D) Repeatable

Question 196

Productivity can be measured by which of the following ratios:

(A) Output:Input
(B) Input:Process time
(C) Output:Process time
(D) Process schedule:Output

Question 197

The customer gap is the difference between the customer view and which of the following?

(A) Producer view as delivered
(B) Producer view as specified
(C) Provider view as delivered
(D) Supplier view as specified

Question 198

What scenario below is least appropriate for the use of check sheets?

(A) To identify the frequency of patterns
(B) Where repeated observations in a single location can occur
(C) Collection of data from a production process
(D) To analyze non-routine events in a process

Question 199

Stratification is a supplemental quality tool best used for what scenario below?

(A) Organizing data from multiple sources
(B) Identifying control limits
(C) Tracking output over time
(D) observing and recording data in a single location

Question 200

Determine the margin of error with a 99% confidence level for a standard deviation of 18 and a sample size of 120.

(A) 0.66
(B) 1.06
(C) 2.55
(D) 4.23

Solution 1

Internal customers are those which purchase, use or support the product or service but also have in interest or play a role in the company. They may be more aware of the benefits or limitations of a product. External are those which are entirely outside of the company.

The answer is **(C)**

Solution 2

Sources of error and variation can be grouped into categories such as the ones listed below:

- Method
- Worker error
- Material defect
- Measurement
- Machine
- Environment

In this example the error comes from the act of measurement.

The answer is **(D)**

Solution 3

Scrap material is that which is excess to the production of a product and will be disposed of. The lack of efficiency in material is a process that can be managed by a company itself and therefore falls under internal failure.

The answer is **(C)**

Solution 4

The total failure cost is a measure of the internal and external failure costs only. Calibration is an appraisal cost which would be a part of the total quality cost but not the total failure cost.

The answer is **(C)**

Solution 5

When developing a cause and effect diagram, it is most often the case to include the following six categories:

- Methods
- Machines
- Measurement
- Material
- Manpower
- Mother Nature

The answer is **(A)**

Solution 6

Flow charts that reach a diverging point where a decision needs to be made are typically represented by a diamond.

The answer is **(B)**

Solution 7

The dependent variable is that which is measured in the experiment in relation to varying independent variables. In this case the samples have varying amounts of water content and the effect on the sample's strength is measured.

The answer is **(D)**

Solution 8

As iterations of the experiment increase, there is a wider set of data and better observations can be made. Therefore, there is a direct relationship between an increase in replications and the accuracy of experimental error.

The answer is **(A)**

Solution 9

The main effect is calculated by subtracting the average of the factor at the high level from the average at the low level and dividing it by two:

$$Main\ Effect = \frac{F_+ - F_-}{2}$$

In this example, we are concerned with the weight factor which has a high level of 4 and a low level of 2:

$$W_+ = \frac{4.2 + 4.3 + 4.6}{3} = 4.367$$

$$W_- = \frac{2.0 + 2.1 + 2.4}{3} = 2.167$$

Then to get the main effect:

$$Main\ Effect = \frac{W_+ - W_-}{2} = \frac{4.367 - 2.167}{2} = 1.1$$

The answer is **(A)**

Solution 10

Discrete distributions are those which are based on counting data and not data measured on a continuous scale such as the standard normal distribution.

The answer is **(C)**

Solution 11

When determining the effect from interactions between two factors the following rules apply:

- If the signs are the same, the interaction is positive
- If the signs are opposite, the interactions are negative.

Therefore, the chart becomes:

Run	A	B	C	AxB	AxC
1	+	+	+	+	+
2	+	+	-	+	-
3	+	-	-	-	-
4	-	+	+	-	-
5	-	-	+	+	-
6	-	-	-	+	+

The answer is **(C)**

Solution 12

A fractional factorial design does not include all possible combinations of factors. Decisions need to be made for what is to be omitted. This is not the ideal scenario in most cases because data is being omitted leading to a smaller sample size. Some of the drawbacks include lower quality of results and the potential for confounded factors. These are factors which because of incomplete data, have the same main effect formula. A positive effect is because of decreased runs, less material is needed and less time.

The answer is **(B)**

Solution 13

A problem definition should include all of the following:

- Problem statement
- Solution Criteria
- Quality cost

At this point the analysis has not been performed and results are not available.

The answer is **(A)**

Solution 14

The 5S tool has 5 steps which are:

- Sort
- Set
- Scrub
- Standardize
- Sustain

The answer is **(D)**

Solution 15

The Plan-Do-Check-Act (PDCA) is a continuous cycle and is repeated infinitely. Therefore, there is no completion of the cycle.

The answer is **(D)**

Solution 16

Observing other organizations is a common way to learn how to increase productivity. This is called benchmarking.

The answer is **(A)**

Solution 17

Control charts are able to track data by plotting points over a period of time.

The answer is **(D)**

Solution 18

This requires the understanding of the different terms used to describe subjects under consideration. A population is the group of all items within the scope of the experiment or project. A sample is a part of the population from which data is collected as a representation of the population. Universe is a less common term which is used the same as population.

The answer is **(C)**

Solution 19

The mean for any standard normal distribution curve is zero since it is symmetrical about the y-axis.

The answer is **(B)**

Solution 20

The Poisson distribution can be used to determine the probability of an occurrence for a specific scenario given an average probability. The equation for the probability (P) of a specific occurrence (x) is:

$$P(x) = \frac{e^{-\bar{c}}\bar{c}^x}{x!}$$

$\bar{c}$ = Average number of defects
X = Number of defects for the specific occurrence

Therefore, in this scenario we have an average of 1.2 defective pieces and are looking for the probability of 2 for a given instance. The equation becomes:

$$P(x) = \frac{e^{-1.2}1.2^2}{2!} = 0.22 = 22\%$$

The answer is **(B)**

Solution 21

The number of combinations for a specific number from a specific set can be determined from the following equation:

$$\frac{n!}{(n-x)!\,x!}$$

Where x is the number of objects and n is the number of objects in the pool:

$$\frac{6!}{(6-3)!\,3!} = \frac{720}{(6)(6)} = 20$$

The answer is **(C)**

Solution 22

For a binomial distribution, the mean is the product of the number of trials (or sample size) and the probability of success:

$$\mu = np = 10(0.25) = 2.5$$

The answer is **(A)**

Solution 23

The dispersion is an indication of how far numbers in the set fall away from the mean. The range, variance and standard deviation all are an indication of this.

The answer is **(B)**

Solution 24

The lower the standard deviation, the more tightly grouped together the values are and therefore is more reliable with less variation.

The answer is **(A)**

Solution 25

For 90% the one-tailed is 1 − 0.9 = 0.1. To get the two-tailed, divide by two 0.1/2 = 0.05.

The answer is **(B)**

Solution 26

The upper and lower limits can be determined by taking the mean and adding or subtracting the margin of error. In this scenario:

Upper Limit = 15,400 + 260 = 15,660

Lower Limit = 15,400 - 260 = 15,140

The answer is **(A)**

Solution 27

Specification limits impose value boundaries based on predetermined numbers that are not founded on a statistical basis. Therefore, they cannot be used to identify a change in process. This is determined by control limits which are based on data.

The answer is **(B)**

Solution 28

There are trends which indicate process change. They include:

- Nonrandom patterns
- Points below or above the upper and lower limits
- Seven successive points above or below the mean

The answer is **(C)**

Solution 29

If defectives are being counted, the appropriate chart is the p or np. With a constant sample size, the np chart is used.

The answer is **(B)**

Solution 30

For a u-chart the control limits are determined by the following:

$$\bar{u} \pm 3 \sqrt{\frac{\bar{u}}{\bar{n}}}$$

$\bar{u}$ = average defect fraction
$\bar{n}$ = average sample size

$$\bar{u} = \frac{0.05 + 0.044 + 0.059 + 0.056 + 0.070}{5} = 0.0558$$

$$\bar{n} = \frac{60 + 45 + 51 + 72 + 71}{5} = 59.8$$

$$UCL = 0.0558 + 3 \sqrt{\frac{0.0558}{59.8}} = 0.147$$

The answer is **(D)**

Solution 31

For centered processes, C_p and C_{pk} have the same value. For non-centered, C_p is greater than C_{pk}.

The answer is **(C)**

Solution 32

Mechanical types of instruments do not have a separate means of measuring the specimen and typically rely on probes that will make contact with a surface.

The answer is **(A)**

Solution 33

There is a difference between errors that occur as isolated incidents and those which are due to the process. The systematic errors are inherent and will often be repeated. There is a source cause that needs to be addressed. Isolated errors are due to issues such as human error and are difficult to prevent. A single reading on a tape measure can be read incorrectly and is not often a sign of a systematic problem.

The answer is **(A)**

Solution 34

Precision and accuracy are commonly confused. Accuracy is the ability to consistently hit a target. Precision is a consistency in outcome regardless of the target. In this case we have a number of shots all relatively close, but none have hit the target of 200 yards. Therefore, it can best be described as precise but not accurate.

The answer is **(B)**

Solution 35

A sine bar is used to measure angular dimensions.

The answer is **(B)**

Solution 36

Working gauges are those used by inspectors or operators during the production process.

The answer is **(B)**

Solution 37

The working tolerance is the part tolerance times the working gauge tolerance:

0.2(0.05) = 0.01 mm

The answer is **(A)**

Solution 38

Fitness-for-use relates specifically to providing a product that meets the needs of the customer in way that they will actually use it.

The answer is **(C)**

Solution 39

There are five basic uses of templates which includes:

1. Inspection of an outside radius tangent to two perpendicular planes
2. Inspection of an inside radius tangent to two perpendicular planes
3. Inspection of a groove
4. Inspection of a ridge segment
5. Inspection of roundness and diameter of a shaft

The answer is **(D)**

Solution 40

Gauge block materials need to be chosen based on factors such as environment, cost, temperature longevity and others. Of the materials listed, steel would be the most vulnerable to corrosion and would have a strong effect on the length of time the block will be useful.

The answer is **(B)**

Solution 41

There are four different types of air pressure gauges:

1. Back-pressure
2. Differential
3. Venturi
4. Flow

The answer is **(B)**

Solution 42

The Rockwell and the Brinell tests measure hardness by imposing a force from on a material and measuring the indentation. The difference is in the material used. The Brinell uses a steel or carbide ball 10 mm in diameter whereas the Rockwell uses a diamond cone.

The answer is **(C)**

Solution 43

The lay is the direction of the predominant surface pattern. ANSI provides symbols to indicate the direction of the lay:

= Parallel
⊥ Perpendicular
X Angular
M Multidirectional
C Circular
R Radial

The answer is **(A)**

Solution 44

The ten-point height (R_z) is determined by taking the average distance between the 5 peaks and 5 valleys within a sampling length.

The answer is **(C)**

Solution 45

The removal of a unit from service should be determined in advance by a database kept by the customers. The unit can then be sent for recalibration. This protects against worn equipment being used.

The answer is **(A)**

Solution 46

A calibration interval may be based on time, regulatory oversight, importance, manufacturer recommendations, or historical performance.

The answer is **(D)**

Solution 47

There are levels of standards so that in the case of a discrepancy, the higher one will govern. The levels from highest to lowest are:

1. International
2. National
3. Primary
4. Secondary
5. Reference
6. Working
7. Transfer

The answer is **(D)**

Solution 48

The maximum material condition (MMC) maximizes the material used and the least material condition (LMC) minimizes it. Therefore, the MMC indicates the maximum shaft and minimum hole and the LMC the minimum shaft and maximum hole.

The answer is **(C)**

Solution 49

Direct measurements are those which measure the desired value directly. An indirect measurement measures something other than a property directly associated with the object in question. For example, the optical flat measures light waves can then be used to determine the desired measurement.

The answer is **(C)**

Solution 50

It is common practice to use the rule of ten when selecting a measurement instrument related to the tolerance of the dimension to be measured. To determine the increment of measure, divide the tolerance by ten: 0.25/10 = 0.025"

The answer is **(B)**

Solution 51

The test uncertainty ratio can be determined from the following equation:

$$TUR = \frac{USL - LSL}{2u} = \frac{0.012 - 0.008}{2(0.0002)} = 10$$

The answer is **(D)**

Solution 52

Uses the conversion to calculate mass from lbs: (110 lbs)(0.4536 kg/lb) = 49.9 kg

The answer is (B)

Solution 53

The gauge R&R can be determined from the following:

$$Gauge\ R\&R = \sqrt{Repeatability^2 + Reproducibility^2} = \sqrt{7.25^2 + 0.75^2} = 7.3$$

The answer is (C)

Solution 54

In this scenario the inspection is in place to ensure the process is working properly up font and then can continue reliably without further investigation. This is known as a verification inspection.

The answer is (C)

Solution 55

Traceability of products can be divided into two different categories. Internal traceability monitors the movement of products within a specific area or manufacturer in a whole supply chain. Chain traceability is the monitoring of products throughout its life moving through multiple processes.

The answer is (A)

Solution 56

The Certificate of Analysis (COA) is used to ensure the material or product achieves a minimum standard set forth by the customer. The Certificate of Conformance (COC) provides documentation of achieving certain specification requirements.

The answer is **(C)**

Solution 57

Nondestructive testing as the name implies allows enhanced inspection of a material without the risk of harm done to the product. Of those listed, X-ray techniques provide displays of the internal features of a part and are not best suited for external defects.

The answer is **(C)**

Solution 58

There are three allowable types of sampling:

1. Single
2. Double
3. Multiple

The answer is **(A)**

Solution 59

The average outgoing quality is calculated by:

$$AOQ = P_a p \left[1 - \frac{Sample\ size}{Lot\ size}\right] = (0.954)(0.01)\left[1 - \frac{20}{200}\right] = 0.0086$$

The answer is **(A)**

Solution 60

Primary quality audits are divided into three types: System, Process and product. They are classified by the extent of the scope and depth of review. The system audit involves reviewing all parts of a system including different processes and personnel. Due to the larger scope, the depth of investigation is less. Process audits identify single processes within the system and provide a narrower scope but an increased depth of analysis. The smallest scope and most detailed audit is the product audit which is focused on a specific product or service.

The answer is **(C)**

Solution 61

There are two risks associated with acceptance sampling plans: producer and consumer. The producer risk is related the lot being rejected while still falling within the acceptable quality level (AQL). The consumer risk is related to acceptance of a lot that has a quality level equal to the lot tolerance percent defective.

The answer is **(C)**

Solution 62

Lot tolerance percent defective is the poorest quality an individual lot can be accepted. It is common in sampling plans to define the LTPD as the percent defective having a 10% probability of acceptance.

The answer is **(B)**

Solution 63

The Material Review Board is responsible for reviewing nonconforming products that have already been segregated and determine the most appropriate disposition. They also shall use the observations and data to try to identify root causes of non-conformance.

The answer is **(D)**

Solution 64

The basic components of an audit include:

- Scope
- Preparation
- Performance
- Documentation
- Closure

The answer is **(C)**

Solution 65

A first-party audit is performed by the client requesting the audit itself. Therefore, it proceeds as an internal audit.

The answer is **(A)**

Solution 66

The preparation component is associated with ensuring the process and personnel are prepared to begin the process. Collecting data does not yet enter into this stage.

The answer is **(D)**

Solution 67

Of the items listed, checklists, tracing and flowcharts are used to identify the steps of a process and proper implementation. Sampling plans are used for quality control measures to inspect in an efficient manner but do not identify process steps.

The answer is **(B)**

Solution 68

The risk priority number is calculated by multiplying the three components together:

RPN = 2 x 4 x 6 = 48

The answer is **(A)**

Solution 69

The first action should be to stop any further damage or repercussions from the issue at hand. Therefore, elimination of the occurrence should be the priority.

The answer is **(A)**

Solution 70

Corrective action is taken to identify the source of a problem and implement a solution to ensure it is not repeated. The elements consist of:

- Identify the problem
- Take temporary action to contain the problem if needed
- Assign personnel to investigate
- Implement the solution
- Monitor the solution to verify effectiveness
- Verify the effectiveness

The answer is **(A)**

Solution 71

Preventive action is used to anticipate a source of concern and to take steps to prevent it from happening. It is not reactionary and therefore a temporary solution that may be needed for corrective action is not often used.

The answer is **(B)**

Solution 72

A process that has a low opportunity for observation will result in more time needed to obtain the proper amount of data.

The answer is **(A)**

Solution 73

Barrier analysis is the evaluation of the point in a process that is holding up or negatively effecting the effectiveness of the process.

The answer is **(B)**

Solution 74

Preventative action is done in anticipation of a potential detrimental situation. This is not in response to something that has occurred which needs correcting.

The answer is **(B)**

Solution 75

The intent of a cause and effect diagram is to find the root cause of a problem and does not likely need to track costs or time.

The answer is **(C)**

Solution 76

Stratification is the organization of data into specific groups to gain clarity on the information as a whole. This is well suited for tools that provide visual representations of data which is not an aspect of fishbone diagrams.

The answer is **(D)**

Solution 77

The PH scale goes from acidic at the lowest number to basic at its highest number.

The answer is **(A)**

Solution 78

The Oscilloscope has the ability to measure voltages as a function of time.

The answer is **(A)**

Solution 79

Comparative is using a master item to use as a reference for the determination of acceptance.

The answer is **(C)**

Solution 80

C_{Pk}/P_{Pk} shifts the center of the distribution away from the mean. This means the mean is not centered between the limits.

The answer is **(D)**

Solution 81

Poor quality costs are those which would not exist if the process or product was perfect or flawless. Warranties are based on anticipating failure and would not be needed if the product had a 100% success rate.

The answer is **(A)**

Solution 82

The Kano model provides three levels of customer expectations:

- Expected: The minimum level of quality anticipated by the customer
- Normal: A reasonable level of expectations beyond the minimum
- Exciting: Features that go beyond expectations and are desirable for the customer

The answer is **(C)**

Solution 83

An affinity diagram is used to organize thoughts into related groups.

The answer is **(A)**

Solution 84

A house of quality is a tool designed to target the customers needs and implement them into the process.

The answer is **(B)**

Solution 85

In the Just-in-time model, there is never an excess of products stocked. This would reduce needed space and costs associated with storage. Products are only produced to meet actual demands. The success is then based greatly on the ability to predict the demand accurately.

The answer is **(C)**

Solution 86

A stable process does not have a tolerance for any special-cause variation and therefore zero are allowed.

The answer is **(A)**

Solution 87

Angles that are not shown but appear to be at a right angle, 180°, or 270° can be assumed and do not need to be dimensioned on the plans.

The answer is (A)

Solution 88

The standard baseline temperature is 68° F. Therefore, any variation in the temperature for the actual material in question must be adjusted for the effects.

The answer is (C)

Solution 89

The customer can receive 0.003 gallons less per gallon. Therefore, if the customer purchases 16 gallons, the minimum is:

$$16 - 0.003(16) = 15.952 \; gallons$$

The answer is (B)

Solution 90

Pyrometers use the following equation to determine temperature:

$$j = \varepsilon \sigma T^4$$

T = Temperature
j = Thermal radiation
$\varepsilon = $ Emissivity
$\sigma = $ Constant of proportionality

The answer is (B)

Solution 91

The binomial distribution is a representation of two outcomes such as accepted or not accepted.

The answer is **(C)**

Solution 92

A switching plan may adapt to the quality of the output by changing the requirements of inspection to less or more stringent. These include normal, reduced, and tightened but there is no auxiliary inspection.

The answer is **(D)**

Solution 93

A single sample plan relies on taking a large number of samples to represent the batch. This is due to only having one iteration of the determination of acceptance

The answer is **(A)**

Solution 94

Multiple sampling plans provide a set number of iterations that the test may go through. The lot can be accepted or rejected during any iteration throughout the process based on the specified acceptance number and rejection number.

The answer is **(C)**

Solution 95

The total cost of quality is the sum of the costs for prevention, appraisal, internal and external failure. Therefore, Cost of Quality = 4 + 8 + 9 + 6 = $27 million

The answer is **(C)**

Solution 96

Internal customers while being able to benefit from a product, may also in certain circumstances be the end user.

The answer is **(D)**

Solution 97

Transcendent view is pursuing a product with superiority and high standards.

The answer is **(C)**

Solution 98

Quality control is more focused on the product itself and ensuring there are no defects present. Quality assurance is more process driven and focuses on prevention and ensuring the proper steps are taken.

The answer is **(A)**

Solution 99

Statistical Quality Control (SQC) is concerned with the output of process and therefore the dependent variables.

The answer is **(B)**

Solution 100

The choice of charts is determined by the number of subgroups.

The answer is **(C)**

Solution 101

A project is bound by three constraints that must be balanced. If one of the three changes it will affect the others. These are cost, scope and schedule.

The answer is **(B)**

Solution 102

Fitness-for-use is the effectiveness in design or process that fits the needs of the targeted consumer. Of the options listed, gathering information directly from consumers and making changes related to this information is mostly related to increasing fitness-for-use of a product.

The answer is **(B)**

Solution 103

Total failure costs are the combination of internal and external failure costs. Appraisal and prevention costs are two of the categories for quality costs but do not factor in to total failure costs.

The answer is **(C)**

Solution 104

The 8D steps include:

- Form a team
- Define the problem
- Develop containment plan
- Identify the root cause
- Choose and verify permanent corrections
- Implement Permanent corrective actions
- Take preventive measures
- Congratulate the team

The answer is **(D)**

Solution 105

Quality Function Deployment is a method for connecting the concerns and interests of the consumer with the product. Therefore, the initial step is to identify the requirements for the customer.

The answer is **(B)**

Solution 106

The independent variable is that which is isolated in the experiment to measure its effect on the dependent variable as it changes. In this case the samples have consistent properties of cement and aggregate but vary in regard to the water content.

The answer is **(C)**

Solution 107

The main effect is a representation of the influence a factor has on a quality characteristic. The further away from zero, the greater indication of influence.

The answer is **(B)**

Solution 108

When determining the effect from interactions between three factors the following rules apply:

- If there is an even number of negatives, the interaction is positive
- If there is an odd number of negatives, the interaction is negative
- All positives are positive, and all negatives are negative

Therefore, the chart becomes:

Run	A	B	C	AxBxC
1	+	+	+	+
2	+	+	-	-
3	+	-	-	+
4	-	+	+	-
5	-	-	+	+
6	-	-	-	-

The answer is **(B)**

Solution 109

The equation for the number of possible runs is:

$$n = L^F$$

In this case the number of runs is predetermined by the maximum of 200. Since we are testing strength, thickness, and temperature, there are 3 factors. Then we can solve for the number of levels:

$$200 = L^3; L = 5.84$$

However, levels must be whole numbers so the maximum is rounded down to 5.

The answer is **(D)**

Solution 110

Scope creep is an addition to the original intended scope and is work that is proposed to fall outside of the original product definition. Estimates, changes in process, and errors may occur that alter the production but if the original intention does not change, it is not scope creep.

The answer is **(A)**

Solution 111

Waste is often classified into seven different categories:

- Transportation
- Inventory
- Motion
- Waiting
- Overproduction
- Over processing
- Defects

In this case the worker is unnecessary having to move far to complete a task. This could be fixed by moving the activation of the machine of possible closer to the workstation.

The answer is **(D)**

Solution 112

In the Kanban system, there is an assembly line with each worker having a specific task involving a specific material. They all have a limited supply of that material. To avoid waste when they run out, a worker called a stocker will be signaled to come over and has an additional supply of all materials so that the depleted one may be resupplied.

The answer is **(C)**

Solution 113

Tools for problem solving can be qualitative or quantitative. Those where number analysis is included are considered hard and those without numbers are soft.

The answer is **(C)**

Solution 114

The code of ethics provides the following main expectations to adhere to the code of ethics:

- Integrity and honesty
- Respect, responsibility and fairness
- Protect proprietary information and avoid conflicts of interest

The answer is **(C)**

Solution 115

The pareto chart is populated by identifying the number of occurrences for different sources. This is useful for the application of identifying the number of times a product fails due to a specific defect.

The answer is **(A)**

Solution 116

The probability of this situation is cumulative of the three possibilities. Therefore:

P = 0.0029 + 0.0161 + 0.0537 = 0.0727

The answer is **(C)**

Solution 117

The area under a standard normal distribution is equal to one square unit. In this case the unit is 0.01 in and therefore the area is 0.01 in².

The answer is **(B)**

Solution 118

For a Poisson distribution, the standard deviation is the square root of the average number of defects:

$$\sigma = \sqrt{\bar{c}} = \sqrt{1.2} = 1.095$$

The answer is **(C)**

Solution 119

The median is the middle point in a set of data points. For an odd number of points, the median is the number with exactly the same amount of points below and above its value. For an even set, the median is the average of the two middle numbers.

The answer is **(D)**

Solution 120

Modes are the most common number in a data set. There can be no mode in the case of all singular data points and there can be multiple modes where there are ties. A histogram measures the frequency of points in a set and therefore the mode will be the highest bar occurring most often.

The answer is **(B)**

Solution 121

The population standard deviation is represented by:

$$\sigma = \sqrt{\frac{\Sigma(x-\mu)^2}{N}}$$

The answer is **(C)**

Solution 122

The normalized confidence coefficient for 95% is 1.96.

The answer is **(C)**

Solution 123

The range is the difference between the highest and lowest value:

R = 190 − 98 = 92

The answer is **(A)**

Solution 124

The variance is the average of the squared differences from the mean

The answer is **(A)**

Solution 125

The process mean is the average of all the averages.

Day 1 average: (11 + 21 + 14 + 11)/4 = 14.25

Day 2 average: (6 + 11 + 12 + 13)/4 = 10.5

Day 3 average: (18 + 4 + 10 + 6)/4 = 9.5

Process mean = (14.25 + + 10.5 + 9.5)/3 = 11.42

The answer is **(A)**

Solution 126

The area under the curve that falls within the range of the distribution will determine the probability of a value falling in that range.

The answer is **(A)**

Solution 127

Control limits are indications of the boundaries that values should fall within for the process and are based on statistical data.

The answer is **(A)**

Solution 128

For defects the chart to use is either u or c. U is used for varying sample size and c for constant.

The answer is **(C)**

Solution 129

The individuals and moving range chart is appropriate for a sample size of one.

The answer is **(A)**

Solution 130

Profilometers analyze the properties of surfaces. There are two types:

- Stylus: Uses a probe which contacts the surface to determine properties
- Optical: non-contact using light to obtain surface information

The answer is **(B)**

Solution 131

All points shown stay within the limits but there is a noticeable pattern that develops. The data seems to have three points that increase and then one decrease. Therefore, the points are appearing nonrandom.

The answer is **(C)**

Solution 132

The average outgoing quality is calculated by:

$$AOQ = P_a p \left[1 - \frac{Sample\ size}{Lot\ size} \right]$$

However, if the lot size is considered infinite it reduces to:

$$AOQ = P_a p = 0.981(0.02) = 0.0196$$

The answer is **(C)**

Solution 133

There are three general classes of measurement:

- Legal: measurements intended to ensure compliance with legal standards
- Technical: Intended to ensure compliance with design specifications or functionality
- Scientific: Used to validate theories and scientific hypotheses

The answer is **(C)**

Solution 134

The capability ratio is the inverse of the capability index C_p which is $1/C_p$

The answer is **(B)**

Solution 135

0.25 mm / 25 = 0.01 mm

The answer is **(B)**

Solution 136

A plug gauge is s cylindrical gauge used to inspect the size of a hole.

The answer is **(A)**

Solution 137

There are three types of tolerance settings. Each of which will have some flaws in regard to unwanted results. A unilateral gauge tolerance where the gauge works within the work tolerance will catch all unacceptable products but will also reject some that are actually acceptable.

Second is the use of a bilateral gauge tolerance. This may allow some of both scenarios to be allowed meaning some acceptable products will be rejected and vice versa.

Last is unilateral where the gauge is outside the working tolerance. This will cause some unacceptable parts to be accepted.

The answer is **(B)**

Solution 138

CMM's have a number of advantages including single setups, data recording, digital readouts, and the ability to save progress. However, they do use a single geometrically fixed measuring system.

The answer is **(A)**

Solution 139

Laboratory master blacks which conform to federal accuracy grade 0.5 are accurate to 0.00003 mm per 25 mm.

The answer is **(A)**

Solution 140

As temperature increases or decreases, there is a slight expansion or contraction associated with materials. If the environment is not well controlled, the measuring devices must be able to adapt appropriately with the changes in the materials being measured. Steel is a material that is quick to adapt to a changing environment and is likely to match any calipers or micrometers being used on site. This makes it an appropriate choice for a poorly controlled environment.

The answer is **(A)**

Solution 141

The probe of a multisensory coordinate measuring machine (MSCMM) is the part which is set to the desired location for relative measurements. Different types are more appropriate for different materials and scenarios. There are contact and non-contact probes. In this scenario we have a textured material meaning the surface is uneven and a point of contact may be inconsistent. Laser scanning probes are able to operate at a distance away from the material and pinpoint an exact point of measurement. This makes it better suited for points on a textured surface.

The answer is **(D)**

Solution 142

The average roughness can be obtained by taking the average of the absolute values of the measurements from a specified coordinate system:

$$R = \frac{\sum|y|}{n} = \frac{0.02 + 0.03 + 0.06 + 0.01 + 0.04}{5} = 0.032$$

The answer is **(C)**

Solution 143

Each dark band indicates a change of one-half wavelength between the flat and work surface.

The answer is **(B)**

Solution 144

The sine bar is laid on a flat surface and then height measurements are taken at each end. The angle can then be determined by the following where A and B are the height measurements:

$$Angle = \frac{B - A}{Length\ of\ Bar} = \frac{4.25 - 0.75}{5} = 0.7$$

Then take the inverted sine:

$$\sin^{-1} 0.7 = 44.43°$$

The answer is **(D)**

Solution 145

Requirements for the use and calibration of measuring instruments is provided by ISO 9001. If and instrument has no applicable international or national standards, it still may be used but the basis used for calibration or verification must be recorded.

The answer is **(B)**

Solution 146

A unilateral tolerance only allows variation in one direction. The superscript indicates the allowable tolerance in a positive or negative direction. Option 'A' has a value in the negative direction but zero in the positive direction meaning it can only be acceptable of the measurement is less than 4.00.

The answer is **(A)**

Solution 147

Maximum Material Condition is a callout where the maximum amount of material exists within a dimensional tolerance. It can only be applied to the following symbols:

- Straightness
- Perpendicularity
- Angularity
- Parallelism
- True position

The answer is **(C)**

Solution 148

The symbol that denotes true position is typically a "target" shape with a circle and two lines crossing the middle perpendicularly.

The answer is **(B)**

Solution 149

There are four levels of classifications for defects which are defined by the level of threat to life and economics. They are:

- Critical: Defect may lead to sever injury or catastrophic economic loss
- Serious: Defect may lead to injury or economic loss
- Major: Defect will cause issues effecting the usability of the product
- Minor: Defect may cause minor issues

The answer is **(B)**

Solution 150

Defect and nonconformity are very similar indicating that a product has failed to meet its standards. The key difference however is that nonconformity refers to the failure to satisfy specification requirements. Defects are related to the product failing to satisfy normal usage requirements or fitness for use.

The answer is **(A)**

Solution 151

Significant digits include those both to the left and right of the decimal point. The number after the 7 is 4 and therefore does not get rounded up.

The answer is **(B)**

Solution 152

Pre-process inspection is in place to ensure there are no identifiable issues before a process begins that may be high cost or prohibit the ability of proper future inspection. Inspecting products that are complete despite the high risk of a historical defect rate is a post-process inspection.

The answer is **(B)**

Solution 153

The test accuracy ratio is a measure of the tolerance of the unit under calibration to the accuracy tolerance of the calibration standard. It is common to set a minimum of 4:1 but lower ratios are accepted at times.

The answer is **(C)**

Solution 154

The eddy current testing method uses AC currents to pass through the material. Therefore, if the material is not conductive, the test will fail.

The answer is **(A)**

Solution 155

The average outgoing quality limit defines the maximum percentage of defective outgoing products that are acceptable.

The answer is **(C)**

Solution 156

As per ANSI/ASQ Z1.4-2008 the seven levels of inspection are:

- Normal
- Reduced
- Tightened
- 4 Special levels

The answer is **(A)**

Solution 157

ANSI/ASQ Z1.9-2008 provide three standard sampling plans:

- Variability known
- Variability unknown standard deviation
- Variability unknown range method

The answer is **(B)**

Solution 158

The three primary audit types are:

- System
- Process
- Product

The answer is **(C)**

Solution 159

The special addition rule of probability states that the probability of either A or B occurring is the addition of the probabilities of the A and B. However, this only applies if A and B cannot occur simultaneously.

The answer is **(B)**

Solution 160

For a double sampling plan, there is only two iterations of the test. Therefore, to ensure a decision, the second test sampling plan rejection number is the acceptance number plus 1.

The answer is **(A)**

Solution 161

There are two risks associated with acceptance sampling plans: producer and consumer. The producer risk is related the lot being rejected while still falling within the acceptable quality level (AQL). The consumer risk is related to acceptance of a lot that has a quality level equal to the lot tolerance percent defective.

The answer is **(A)**

Solution 162

Once a product or material has been segregated, an analysis must take place to determine the best course of action. The product may be scrapped or even reinserted at a later time, but this cannot take place immediately without consideration.

The answer is **(A)**

Solution 163

The bolt is determined to be usable and shippable in its current condition without any alterations necessary. Therefore, it can be designated "Use as-is".

The answer is **(A)**

Solution 164

A Second-party audit is performed by an entity is outside of the organization in question, but which also has a business interest in the audit.

The answer is **(B)**

Solution 165

Performance is the execution of the audit and well beyond the identification of the authorization source.

The answer is **(A)**

Solution 166

There is no requirement for the size of the data set.

The answer is **(B)**

Solution 167

The risk priority number is a function three factors:

- Severity
- Occurrence
- Detection

The answer is **(A)**

Solution 168

The RPN is calculated by multiplying the three components together which each have a maximum of 10. Therefore, the maximum is 10 x 10 x 10 = 1000.

The answer is **(D)**

Solution 169

It is acceptable to rate both the failure and the cause. Proper documentation of assumptions is required for consistency.

The answer is **(C)**

Solution 170

An FMEA may be defined by the extent of the scope and can be system, design, process or service delivery.

The answer is **(C)**

Solution 171

Corrective action should eliminate the source of the issue and ensure a permanent solution. However, it is most often appropriate to take immediate action to limit the short-term damage. This may not be a fully developed solution and can be temporary.

The answer is **(D)**

Solution 172

The root cause evaluation is a part of the process for determining the corrective action plan and is not a means of evaluating the effectiveness of the plan.

The answer is **(B)**

Solution 173

Pareto is the method of identifying the tasks which are most impactful and emphasizing their efficiency.

The answer is **(A)**

Solution 174

Corrective action is done in in response to a discovery of something causing issues within a process.

The answer is **(D)**

Solution 175

Corrective action is well thought out to ensure any action taken addresses the root cause of the problem.

The answer is **(D)**

Solution 176

The type of geometry used will have an effect on the ability to identify specific outputs. Of those listed there is no variable geometry.

The answer is **(B)**

Solution 177

PH is a measure of how acidic or basic water is. A neutral rating is where the water is directly in the middle which is a rating of 7.

The answer is **(C)**

Solution 178

The LCR meter is used to measure inductance, capacitance and resistance.

The answer is **(D)**

Solution 179

The capability index is an indication of the number of products that will fall outside of the tolerances within a given process. Most processes prefer to be around $C_P > 1.33$. If the index is less than 1 then the allowable variation is less than the process variation and values will fall outside of the limits.

The answer is **(A)**

Solution 180

The ability of a process to meet the lower limit is determined by C_{Pl}/P_{Pl}

The answer is **(B)**

Solution 181

Good quality is associated with taking steps to prevent errors.

The answer is **(C)**

Solution 182

The Kano model provides different levels of attributes that the customer can react to:

- Expected: the minimum level of expectations. If these are not available, the customer will be dissatisfied but cannot be fully satisfied by this alone
- Normal: requirements that are able to satisfy a customer and push them into being satisfied
- Exciting: those which go beyond what the customer expects and can fully satisfy them

The answer is **(A)**

Solution 183

Quality Function Deployment is focused on identifying customer needs and taking steps to provide them throughout the process.

The answer is **(C)**

Solution 184

The Charpy V-notch tests uses a pendulum to strike a notched specimen. The contact and damage makes it a destructive test.

The answer is **(D)**

Solution 185

A pull system provides goods and services as they are required by the customer.

The answer is **(A)**

Solution 186

Special cause variation is a change in process due to a specific change whether it be environmental or input. The use a a new cement type is a specific change in the variables in the process.

The answer is **(A)**

Solution 187

Many measurements need to have a reference in which to measure to. For instance, a line on its own cannot be perpendicular without a line or plane to refer to. Flatness is an isolated quality that can occur independent of a datum.

The answer is **(A)**

Solution 188

While there are recommended practices and variation depending on the equipment, calibration frequency has no minimum standard. Instead the user shall make a judgement based on a number of factors such as performance history, importance, quality and others.

The answer is **(D)**

Solution 189

Since there is a confidence of 95%, this means that there is a 5% chance that the actual value falls outside of the reading. The normal distribution is symmetrical, with half of the values outside of the range to the high side above 11.05 V and half below 10.95. Therefore, you can take 5%/2 = 2.5% for the values falling below 10.95.

The answer is (B)

Solution 190

Since the material is continuous and has a uniform thickness, the 1-D scanning is most appropriate.

The answer is (A)

Solution 191

Since the capability ratio is the inverse of the capability index, the lower the value, the more stable the process is.

The answer is (C)

Solution 192

Following the rules, the cycle begins at normal and then 0 are rejected with 10 accepted so there is a switch to reduced. Then we have one rejected which moves it back to normal. The we have 1 out of 10 which is not enough to move to tightened and it remains at Normal.

The answer is (B)

Solution 193

Due to the largest number of samples and time, the cost for a single sampling plan is the often the highest.

The answer is **(A)**

Solution 194

Good quality is a function of appraisal and prevention.

The answer is **(C)**

Solution 195

Special cause variation is caused by a known factor that is non-random and therefore are also known as assignable.

The answer is **(A)**

Solution 196

The measure of productivity is the output to the input.

The answer is **(A)**

Solution 197

The customer gap is what the customer views they have received and what the producer is providing.

The answer is **(A)**

Solution 198

Check sheets are most appropriate for single location observations and collection of data. It does not go into deeper analysis of special events.

The answer is **(D)**

Solution 199

Stratification is best suited for condensing and organizing data from multiple sources.

The answer is **(A)**

Solution 200

The margin of error is determined by the following:

$$Margin\ of\ Error = \frac{Z_{\alpha/2}s}{\sqrt{n}} = \frac{2.575(18)}{\sqrt{120}} = 4.23$$

The answer is **(D)**

Answer Key

1	C	41	B	81	A	121	C	161	A
2	D	42	C	82	C	122	C	162	A
3	C	43	A	83	A	123	A	163	A
4	C	44	C	84	B	124	A	164	B
5	A	45	A	85	C	125	A	165	A
6	B	46	D	86	A	126	A	166	B
7	D	47	D	87	A	127	A	167	A
8	A	48	C	88	C	128	C	168	D
9	A	49	C	89	B	129	A	169	C
10	C	50	B	90	B	130	B	170	C
11	C	51	D	91	C	131	C	171	D
12	B	52	B	92	D	132	C	172	B
13	A	53	C	93	A	133	C	173	A
14	D	54	C	94	C	134	B	174	D
15	D	55	A	95	C	135	B	175	D
16	A	56	C	96	D	136	A	176	B
17	D	57	C	97	C	137	B	177	C
18	C	58	A	98	A	138	A	178	D
19	B	59	A	99	B	139	A	179	A
20	B	60	C	100	C	140	A	180	B
21	C	61	C	101	B	141	D	181	C
22	A	62	B	102	B	142	C	182	A
23	B	63	D	103	C	143	B	183	C
24	A	64	C	104	D	144	D	184	D
25	B	65	A	105	B	145	B	185	A
26	A	66	D	106	C	146	A	186	A
27	B	67	B	107	B	147	C	187	A
28	C	68	A	108	B	148	B	188	D
29	B	69	A	109	D	149	B	189	B
30	D	70	A	110	A	150	A	190	A
31	C	71	B	111	D	151	B	191	C
32	A	72	A	112	C	152	B	192	B
33	A	73	B	113	C	153	C	193	A
34	B	74	B	114	C	154	A	194	C
35	B	75	C	115	A	155	C	195	A
36	B	76	D	116	C	156	A	196	A
37	A	77	A	117	B	157	B	197	A
38	C	78	A	118	C	158	C	198	D
39	D	79	C	119	D	159	B	199	A
40	B	80	D	120	B	160	A	200	D

Reference

The Certified Quality Technician Handbook. H. Fred Walker, Donald W. Benbow, Ahmad K. Elshennawy. Third Edition